A DECLASSIFIED GUIDE
TO **SUBVERTING FASCISM** THROUGH
WEAPONIZED INCOMPETENCE

SIMPLE SABOTAGE FIELD MANUAL

UNITED STATES OFFI
OF STRATEG

BLUES
BOO

SIMPLE SABOTAGE FIELD MANUAL

Published by:
Bluestone Books
www.bluestonebooks.co

ISBN 978-1-965636-28-2 (paperback)
ISBN 978-1-965636-29-9 (e-book)

Printed in Canada
First Edition: 2025

10 9 8 7 6 5 4 3 2 1

CONTENTS

OFFICE OF
STRATEGIC SERVICES

DECLASSIFIE

A NOTE FROM THE PUBLISHER

This book comprises the original World War II–era "Simple Sabotage Field Manual"—a historical brief issued by the United States Office of Strategic Services (the predecessor to the Central Intelligence Agency). It was designed to guide resistance and subversive activities during a time of global conflict. Although its strategies were conceived for a very different era, the manual remains a fascinating window into the art of subtle disruption and the creative problem-solving that emerges under pressure. Today, its insights invite us to reflect on the dynamics of strategy, resilience, and the power of unconventional thinking in both historical and modern contexts. We are proud to present this document in its original, unedited form, preserving its historical authenticity for contemporary readers.

OSS REPRODUCTION BRANCH
SIMPLE SABOTAGE FIELD MANUAL
Strategic Services
(Provisional)

STRATEGIC SERVICES FIELD MANUAL No. 3
Office of Strategic Services
Washington, D. C.

17 January 1944

This Simple Sabotage Field Manual Strategic Services (Provisional) is published for the information and guidance of all concerned and will be used as the basic doctrine for Strategic Services training for this subject.

The contents of this Manual should be carefully controlled and should not be allowed to come into unauthorized hands.

The instructions may be placed in separate pamphlets or leaflets according to categories of operations but should be distributed with care and not broadly.

They should be used as a basis of radio broadcasts only for local and special cases and as directed by the theater commander.

AR 380-5, pertaining to handling of secret documents, will be complied with in the handling of this Manual.

William J. Donovan

• 1 •

INTRODUCTION

The purpose of this paper is to characterize simple sabotage, to outline its possible effects, and to present suggestions for inciting and executing it.

Sabotage varies from highly technical coup de main acts that require detailed planning and the use of specially trained operatives, to innumerable simple acts which the ordinary individual citizen-saboteur can perform. This paper is primarily concerned with the latter type. Simple sabotage does not require specially prepared tools or equipment; it is executed by an ordinary citizen who may or may not act individually and without the necessity for active connection with an organized group; and it is carried out in such a way as to involve a minimum danger of injury, detection, and reprisal.

Where destruction is involved, the weapons of the citizen-saboteur are salt, nails, candles, pebbles, thread, or any other materials he might normally be expected to possess as a householder or as a worker in his particular occupation. His arsenal is the kitchen shelf, the trash pile, his own usual kit of tools and supplies. The targets of his sabotage are usually objects to which he has normal and inconspicuous access in everyday life.

A second type of simple sabotage requires no destructive tools whatsoever and produces physical damage, if any, by highly indirect means. It is based on universal opportunities to make faulty decisions, to adopt a noncooperative attitude, and to induce others to follow suit. Making a faulty decision may be simply a matter of placing tools in one spot instead of another. A noncooperative attitude may involve nothing more than creating an unpleasant situation among one's fellow workers, engaging in bickerings, or displaying surliness and stupidity.

This type of activity, sometimes referred to as the "human element," is frequently responsible for accidents, delays, and general obstruction even under normal conditions. The potential saboteur should discover what types of faulty decisions and operations are normally found in this kind of work and should then devise his sabotage so as to enlarge that "margin for error."

2

POSSIBLE EFFECTS

Acts of simple sabotage are occurring throughout Europe. An effort should be made to add to their efficiency, lessen their detectability, and increase their number. Acts of simple sabotage, multiplied by thousands of citizen-saboteurs, can be an effective weapon against the enemy. Slashing tires, draining fuel tanks, starting fires, starting arguments, acting stupidly, short-circuiting electric systems, and abrading machine parts will waste materials, manpower, and time. Occurring on a wide scale, simple sabotage will be a constant and tangible drag on the war effort of the enemy.

Simple sabotage may also have secondary results of more or less value. Widespread practice of simple sabotage will harass and demoralize enemy administrators and police. Further, success may embolden the citizen-saboteur eventually to find colleagues who can assist him in sabotage of greater dimensions. Finally, the very practice of simple sabotage by natives in enemy or occupied territory may make these individuals identify themselves actively with the United Nations war effort, and encourage them to assist openly in periods of Allied invasion and occupation.

• 3 •

MOTIVATING THE SABOTEUR

To incite the citizen to the active practice of simple sabotage and to keep him practicing that sabotage over sustained periods is a special problem.

Simple sabotage is often an act which the citizen performs according to his own initiative and inclination. Acts of destruction do not bring him any personal gain and may be completely foreign to his habitually conservationist attitude toward materials and tools. Purposeful stupidity is contrary to human nature. He frequently needs pressure, stimulation or assurance, and information and suggestions regarding feasible methods of simple sabotage.

PERSONAL MOTIVES

(A) The ordinary citizen very probably has no immediate personal motive for committing simple sabotage. Instead, he must be made to anticipate indirect personal gain, such as might come with enemy evacuation or destruction of the ruling government group. Gains should be stated as specifically as possible for the area addressed: simple sabotage will hasten the day when Commissioner X and his deputies Y and

Z will be thrown out, when particularly obnoxious decrees and restrictions will be abolished, when food will arrive, and so on. Abstract verbalizations about personal liberty, freedom of the press, and so on, will not be convincing in most parts of the world. In many areas they will not even be comprehensible.

(B) Since the effect of his own acts is limited, the saboteur may become discouraged unless he feels that he is a member of a large, though unseen, group of saboteurs operating against the enemy or the government of his own country and elsewhere. This can be conveyed indirectly: suggestions which he reads and hears can include observations that a particular technique has been successful in this or that district. Even if the technique is not applicable to his surroundings, another's success will encourage him to attempt similar acts. It also can be conveyed directly: statements praising the effectiveness of simple sabotage can be contrived which will be published by white radio, freedom stations, and the subversive press. Estimates of the proportion of the population engaged in sabotage can be disseminated. Instances of successful sabotage already are being broadcast by white radio and freedom stations, and this should be continued and expanded where compatible with security.

(C) More important than **(A)** or **(B)** would be to create a situation in which the citizen-saboteur acquires a sense of responsibility and begins to educate others in simple sabotage.

ENCOURAGING DESTRUCTIVENESS

It should be pointed out to the saboteur, where the circumstances are suitable, that he is acting in self-defense against the enemy, or retaliating against the enemy for other acts of destruction. A reasonable amount of humor in the presentation of suggestions for simple sabotage will relax tensions of fear.

(A) The saboteur may have to reverse his thinking, and he should be told this in so many words. Where he formerly thought of keeping his tools sharp, he should now let them grow dull; surfaces that formerly were lubricated now should be sanded; normally diligent, he should now be lazy and careless; and so on. Once he is encouraged to think backwards about himself and the objects of his everyday life, the saboteur will see many opportunities in his immediate environment which cannot possibly be seen from a distance. A state of mind should be encouraged that anything can be sabotaged.

(B) Among the potential citizen-saboteurs who are to engage in physical destruction, two extreme types may be distinguished. On the one hand, there is the man who is not technically trained and employed. This man needs specific suggestions as to what he can and should destroy as well as details regarding the tools by means of which destruction is accomplished.

(C) At the other extreme is the man who is a technician, such as a lathe operator or an automobile mechanic. Presumably this man would be able to devise methods of simple sabotage which would be appropriate to his own facilities. However, this man needs to be stimulated to reorient his thinking in the direction of destruction. Specific examples, which need not be from his own field, should accomplish this.

(D) Various media may be used to disseminate suggestions and information regarding simple sabotage. Among the media which may be used, as the immediate situation dictates, are: freedom stations or radio false (unreadable) broadcasts or leaflets may be directed toward specific geographic or occupational areas, or they may be general in scope. Finally, agents may be trained in the art of simple sabotage in anticipation of a time when they may be able to communicate this information directly.

SAFETY MEASURES

(A) The amount of activity carried out by the saboteur will be governed not only by the number of opportunities he sees, but also by the amount of danger he feels. Bad news travels fast, and simple sabotage will be discouraged if too many simple saboteurs are arrested.

(B) It should not be difficult to prepare leaflets and other media for the saboteur about the choice of weapons, time, and targets which will insure the saboteur against detection and retaliation. Among such suggestions might be the following:

(1) Use materials which appear to be innocent. A knife or a nail file can be carried normally on your person; either is a multipurpose instrument for creating damage. Matches, pebbles, hair, salt, nails, and dozens of other destructive agents can be carried or kept in your living quarters without exciting any suspicion whatever. If you are a worker in a particular trade or industry, you can easily carry and keep such things as wrenches, hammers, emery paper, and the like.

(2) Try to commit acts for which large numbers of people could be responsible. For instance, if you blow out the wiring in a factory at a central fire box, almost anyone could have done it. On-the-street sabotage after dark, such as you might be able to carry out against a military car or truck, is another example of an act for which it would be impossible to blame you.

(3) Do not be afraid to commit acts for which you might be blamed directly, as long as you do so rarely, and as long as you have a plausible excuse: you dropped your wrench across an electric circuit because an air raid had kept you up the night before and you were half-dozing at work. Always be profuse in your apologies. Frequently you can "get away" with such acts under the cover of pretending stupidity, ignorance, over-caution, fear of being suspected of sabotage, or weakness and dullness due to undernourishment.

(4) After you have committed an act of easy sabotage, resist any temptation to wait around and see what happens. Loiterers arouse suspicion. Of course, there are circumstances when it would be suspicious for you to leave. If you commit sabotage on your job, you should naturally stay at your work.

4

TOOLS, TARGETS, AND TIMING

The citizen-saboteur cannot be closely controlled. Nor is it reasonable to expect that simple sabotage can be precisely concentrated on specific types of targets according to the requirements of a concrete military situation. Attempts to control simple sabotage according to developing military factors, moreover, might provide the enemy with intelligence of more or less value in anticipating the date and area of notably intensified or notably slackened military activity.

Sabotage suggestions, of course, should be adapted to fit the area where they are to be practiced. Target priorities for general types of situations likewise can be specified for emphasis at the proper time by the underground press, freedom stations, and cooperating propaganda.

UNDER GENERAL CONDITIONS

(A) Simple sabotage is more than malicious mischief, and it should always consist of acts whose results will be detrimental to the materials and manpower of the enemy.

(B) The saboteur should be ingenious in using his everyday equipment. All sorts of weapons will present themselves if he looks at his surroundings in a different light. For example, emery dust—a powerful weapon—may at first seem unobtainable, but if the saboteur were to pulverize an emery knife sharpener or emery wheel with a hammer, he would find himself with a plentiful supply.

(C) The saboteur should never attack targets beyond his capacity or the capacity of his instruments. An inexperienced person should not, for example, attempt to use explosives, but should confine himself to the use of matches or other familiar weapons.

(D) The saboteur should try to damage only objects and materials known to be in use by the enemy or to be destined for early use by the enemy. It will be safe for him to assume that almost any product of heavy industry is destined for enemy use, and that the most efficient fuels and lubricants also are destined for enemy use. Without special knowledge, however, it would be undesirable for him to attempt destruction of food crops or food products.

(E) Although the citizen-saboteur may rarely have access to military objects, he should give these preference above all others.

PRIOR TO A MILITARY OFFENSIVE

During periods which are quiescent in a military sense, such emphasis as can be given to simple sabotage might well center on industrial production, to lessen the flow of materials and equipment to the enemy. Slashing a rubber tire on an Army truck may be an act of value; spoiling a batch of rubber in the production plant is an act of still more value.

DURING A MILITARY OFFENSIVE

(A) Most significant sabotage for an area which is, or is soon destined to be, a theater of combat operations is that whose effects will be direct and immediate. Even if the effects are relatively minor and localized, this type of sabotage is to be preferred to activities whose effects, while widespread, are indirect and delayed.

(1) The saboteur should be encouraged to attack transportation facilities of all kinds. Among such facilities are roads, railroads, automobiles, trucks, motorcycles, bicycles, trains, and trams.

(2) Any communications facilities which can be used by the authorities to transmit instructions or morale material should be the objects of simple sabotage. These include telephone, telegraph, and power systems, radio, newspapers, placards, and public notices.

(3) Critical materials, valuable in themselves or necessary to the efficient functioning of transportation and communication, also should become targets for the citizen-saboteur. These may include oil, gasoline, tires, food, and water.

• 5 •

SPECIFIC SUGGESTIONS FOR SIMPLE SABOTAGE

It will not be possible to evaluate the desirability of simple sabotage in an area without having in mind rather specifically what individual acts and results are embraced by the definition of simple sabotage.

A listing of specific acts follows, classified according to types of target. This list is presented as a growing rather than a complete outline of the methods of simple sabotage. As new techniques are developed, or new fields explored, it will be elaborated and expanded.

BUILDINGS

Warehouses, barracks, offices, hotels, and factory buildings are outstanding targets for simple sabotage. They are extremely susceptible to damage, especially by fire; they offer opportunities to such untrained people as janitors, charwomen, and casual visitors; and, when damaged, they present a relatively large handicap to the enemy.

(A) Fires can be started wherever there is an accumulation of inflammable material. Warehouses are obviously the most promising targets, but incendiary sabotage need not be confined to them alone.

(1) Whenever possible, arrange to have the fire start after you have gone away. Use a candle and paper combination, setting it as close as possible to the inflammable material you want to burn: From a sheet of paper, tear a strip three or four centimeters wide and wrap it around the base of the candle two or three times. Twist more sheets of paper into loose ropes and place them around the base of the candle. When the candle flame reaches the encircling strip, it will be ignited and in turn will ignite the surrounding paper. The size, heat, and duration of the resulting flame will depend on how much paper you use and how much of it you can cramp in a small space.

(2) With a flame of this kind, do not attempt to ignite any but rather inflammable materials, such as cotton sacking. To light more resistant materials, use a candle plus tightly rolled or twisted paper which has been soaked in gasoline. To create a briefer but even hotter flame, put celluloid, such as you might find in an old comb, into a nest of plain or saturated paper which is to be fired by a candle.

(3) To make another type of simple fuse, soak one end of a piece of string in grease. Rub a generous pinch of gunpowder over the inch of string where greasy string meets clean string. Then ignite the clean end of the string. It will burn slowly without a flame (in much the same way that a cigarette burns) until it reaches the grease and gunpowder; it will then flare up suddenly. The grease-treated string will then burn with a flame. The same effect may be achieved by using matches instead of the grease and gunpowder. Run the string over the match heads, taking care that the string is not pressed or knotted. They too will produce a sudden flame. The advantage of this type of fuse is that string burns at a set speed. You can time your fire by the length and thickness of the string you chose.

(4) Use a fuse such as the ones suggested above to start a fire in an office after hours. The destruction of records and other types of documents would be a serious handicap to the enemy.

(5) In basements where waste is kept, janitors should accumulate oily and greasy waste. Such waste sometimes ignites spontaneously, but it can easily be lit with a cigarette or match. If you are a janitor on night duty, you can be the first to report the fire, but don't report it too soon.

(6) A clean factory is not susceptible to fire, but a dirty one is. Workers should be careless with refuse and janitors should be inefficient in cleaning. If enough dirt and trash can be accumulated, an otherwise fireproof building will become inflammable.

(7) Where illuminating gas is used in a room which is vacant at night, shut the windows tightly, turn on the gas, and leave a candle burning in the room, closing the door tightly behind you. After a time, the gas will explode, and a fire may or may not follow.

(B) Water and Miscellaneous

(1) Ruin warehouse stock by setting the automatic sprinkler system to work. You can do this by tapping the sprinkler heads sharply with a hammer or by holding a match under them.

(2) Forget to provide paper in toilets; put tightly rolled paper, hair, and other obstructions in the W. C. Saturate a sponge with a thick starch or sugar solution. Squeeze it tightly into a ball, wrap it with string, and dry. Remove the string when fully dried. The sponge will be in the form of a tight hard ball. Flush down a W. C. or otherwise introduce into a sewer line. The sponge will gradually expand to its normal size and plug the sewage system.

(3) Put a coin beneath a bulb in a public building during the daytime, so that fuses will blow out when lights are turned on at night. The fuses themselves may be rendered ineffective by putting a coin behind them or loading them with heavy wire. Then a shortcircuit may either start a fire, damage transformers, or blow out a central fuse which will interrupt distribution of electricity to a large area.

(4) Jam paper, bits of wood, hairpins, and anything else that will fit into the locks of all unguarded entrances to public buildings.

INDUSTRIAL PRODUCTION: MANUFACTURING

(A) Tools

(1) Let cutting tools grow dull. They will be inefficient, will slow down production, and may damage the materials and parts you use them on.

(2) Leave saws slightly twisted when you are not using them. After a while, they will break when used.

(3) Using a very rapid stroke will wear out a file before its time. So will dragging a file in slow strokes under heavy pressure. Exert pressure on the backward stroke as well as the forward stroke.

(4) Clean files by knocking them against the vise or the workpiece; they are easily broken this way.

(5) Bits and drills will snap under heavy pressure.

(6) You can put a press punch out of order by putting in it more material than it is adjusted for—two blanks instead of one, for example.

(7) Power-driven tools like pneumatic drills, riveters, and so on are never efficient when dirty. Lubrication points and electric contacts can easily be fouled by normal accumulations of dirt or the insertion of foreign matter.

(B) Oil and Lubrication Systems

These are not only vulnerable to easy sabotage, but are critical in every machine with moving parts. Sabotage of oil and lubrication will slow production or stop work entirely at strategic points in industrial processes.

(1) Put metal dust or filings, fine sand, ground glass, emery dust (get it by pounding up an emery knife sharpener), and similar hard, gritty substances directly into lubrication systems. They will scour smooth surfaces, ruining pistons, cylinder walls, shafts, and bearings. They will overheat and stop motors, which will need overhauling, new parts, and extensive repairs. Such materials, if they are used, should be introduced into lubrication systems past any filters which otherwise would strain them out.

(2) You can cause wear on any machine by uncovering a filter system, poking a pencil or any other sharp object through the filter mesh, then covering it up again. Or, if you can dispose of it quickly, simply remove the filter.

(3) If you cannot get at the lubrication system or filter directly, you may be able to lessen the effectiveness of oil by diluting it in storage. In this case, almost any liquid will do which will thin the oil. A small amount of sulphuric acid, varnish, water glass, or linseed oil will be especially effective.

(4) Using a thin oil where a heavy oil is prescribed will break down a machine or heat up a moving shaft so that it will "freeze" and stop.

(5) Put any clogging substance into lubrication systems or, if it will float, into stored oil. Twisted combings of human hair, pieces of string, dead insects, and many other common objects will be effective in stopping or hindering the flow of oil through feed lines and filters.

(6) Under some circumstances, you may be able to destroy oil outright rather than interfere with its effectiveness, by removing stop plugs from lubricating systems or by puncturing the drums and cans in which it is stored.

(C) Cooling Systems

(1) A water cooling system can be put out of commission in a fairly short time, with considerable damage to an engine or motor, if you put into it several pinches of hard grain, such as rice or wheat. They will swell up and choke the circulation of water, and the cooling system will have to be torn down to remove the obstruction. Sawdust or hair may also be used to clog a water cooling system.

(2) If very cold water is quickly introduced into the cooling system of an overheated motor, contraction and considerable strain on the engine housing will result. If you can repeat the treatment a few times, cracking and serious damage will result.

(3) You can ruin the effectiveness of an air cooling system by plugging dirt and waste into intake or exhaust valves. If a belt-run fan is used in the system, make a jagged cut at least halfway through the belt; it will slip and finally part under strain and the motor will overheat.

(D) Gasoline and Oil Fuel Tanks and Fueling Engines

These usually are accessible and easy to open. They afford a very vulnerable target for simple sabotage activities.

(1) Put several pinches of sawdust or hard grain, such as rice or wheat, into the fuel tank of a gasoline engine. The particles will choke a feed line so that the engine will stop. Some time will be required to discover the source of the trouble. Although they will be hard to get, crumbs of natural rubber, such as you might find in old rubber bands and pencil erasers, are also effective.

(2) If you can accumulate sugar, put it in the fuel tank of a gasoline engine. As it burns together with the gasoline, it will turn into a sticky mess, which will completely mire the engine and necessitate extensive cleaning and repair. Honey and molasses are as good as sugar. Try to use about 75–100 grams for each 10 gallons of gasoline.

(3) Other impurities which you can introduce into gasoline will cause rapid engine wear and eventual breakdown. Fine particles of pumice, sand, ground glass, and metal dust can easily be introduced into a gasoline tank. Be sure that the particles are very fine, so that they will be able to pass through the carburetor jet.

(4) Water, urine, wine, or any other simple liquid you can get in reasonably large quantities will dilute gasoline fuel to a point where no combustion will occur in the cylinder and the engine will not move. One pint to 20 gallons of gasoline is sufficient. If saltwater is used, it will cause corrosion and permanent motor damage.

(5) In the case of diesel engines, put low-flashpoint oil into the fuel tank; the engine will not move. If there already is proper oil in the tank when the wrong kind is added, the engine will only limp and sputter along.

(6) Fuel lines to gasoline and oil engines frequently pass over the exhaust pipe. When the machine is at rest, you can stab a small hole in the fuel line and plug the hole with wax. As the engine runs and the exhaust tube becomes hot, the wax will be melted; fuel will drip onto the exhaust and a blaze will start.

(7) If you have access to a room where gasoline is stored, remember that gas vapor accumulating in a closed room will explode after a time if you leave a candle burning in the room. A good deal of evaporation, however, must occur from the gasoline tins into the air of the room. If removal of the tops of the tins does not expose enough gasoline to the air to ensure copious evaporation, you can open lightly constructed tins further with a knife, ice pick, or sharpened nail file. Or puncture a tiny hole in the tank, which will permit gasoline to leak out on the floor. This will greatly increase the rate of evaporation. Before you light your candle, be sure that windows are closed and the room is as airtight as you can make it. If you can see that windows in a neighboring room are opened wide, you have a chance of setting a large fire which will not only destroy the gasoline but anything else nearby; when the gasoline explodes, the doors of the storage room will be blown open and a draft to the neighboring windows will be created, which will whip up a fine conflagration.

(E) Electric Motors

Electric motors (including dynamos) are more restricted than the targets so far discussed. They cannot be sabotaged easily or without risk of injury by unskilled persons who may otherwise have good opportunities for destruction.

(1) Set the rheostat to a high point of resistance in all types of electric motors. They will overheat and catch fire.

(2) Adjust the overload relay to a very high value beyond the capacity of the motor. Then overload the motor to a point where it will overheat and break down.

(3) Remember that dust, dirt, and moisture are enemies of electrical equipment. Spill dust and dirt onto the points where the wires in electric motors connect with terminals, and onto insulating parts. Inefficient transmission of current and, in some cases, short circuits will result. Wet generator motors to produce short circuits.

(4) "Accidentally" bruise the insulation on wire, loosen nuts on connections, and make faulty splices and faulty connections in wiring to waste electric current and reduce the power of electric motors, decrease the power output, or cause short-circuiting in direct-current motors: Loosen or remove commutator holding rings. Sprinkle carbon, graphite, or metal dust on commutators. Put a little grease or oil at the contact points of commutators. Where commutator bars are close together, bridge the gaps between them with metal dust, or sawtooth their edges with a chisel so that the teeth on adjoining bars meet or nearly meet and current can pass from one to the other.

(6) Put a piece of finely grained emery paper half the size of a postage stamp in a place where it will wear away rotating brushes. The emery paper and the motor will be destroyed in the resulting fire.

(7) Sprinkle carbon, graphite, or metal dust on slip rings so that the current will leak or short circuits will occur. When a motor is idle, nick the slip rings with a chisel.

(8) Cause motor stoppage or inefficiency by applying dust mixed with grease to the face of the armature so that it will not make proper contact.

(9) To overheat electric motors, mix sand with heavy grease and smear it between the stator and rotor, or wedge thin metal pieces between them. To prevent the efficient generation of current, put floor sweepings, oil, tar, or paint between them.

(10) In motors using three-phase current, deeply nick one of the lead-in wires with a knife or file when the machine is at rest, or replace one of the three fuses with a blown-out fuse. In the first case, the motor will stop after running a while, and in the second, it will not start.

(F) Transformers

(1) Transformers of the oil-filled type can be put out of commission if you pour water, saltwater, machine-tool coolant, or kerosene into the oil tank.

(2) In air-cooled transformers, block the ventilation by piling debris around the transformer.

(3) In all types of transformers, throw carbon, graphite, or metal dust over the outside bushings and other exposed electrical parts.

(G) Turbines

For the most part, turbines are heavily built, stoutly housed, and difficult to access. Their vulnerability to simple sabotage is very low.

(1) After inspecting or repairing a hydro turbine, fasten the cover insecurely so that it will blow off and flood the plant with water. A loose cover on a steam turbine will cause it to leak and slow down.

(2) In water turbines, insert a large piece of scrap iron in the head of the penstock, just beyond the screening, so that water will carry the damaging material down to the plant equipment.

(3) When the steam line to a turbine is opened for repair, put pieces of scrap iron into it, to be blasted into the turbine machinery when steam is up again.

(4) Create a leak in the line feeding oil to the turbine, so that oil will fall on the hot steam pipe and cause a fire.

(H) Boilers

(1) Reduce the efficiency of steam boilers any way you can. Put too much water in them to make them slow-starting, or keep the fire under them low to keep them inefficient. Let them dry and turn the fire up; they will crack and be ruined. An especially good trick is to keep putting limestone or water containing lime in the boiler; it will deposit lime on the bottom and sides. This deposit will provide very good insulation against heat; after enough of it has collected, the boiler will be completely worthless.

PRODUCTION: METALS

(A) Iron and Steel

(1) Keep blast furnaces in a condition where they must

be frequently shut down for repair. In making fireproof bricks for the inner lining of blast furnaces, put in an extra proportion of tar so that they will wear out quickly and necessitate constant relining.

(2) Make cores for casting so that they are filled with air bubbles and an imperfect cast results.

(3) See that the core in a mold is not properly supported, so that the core gives way or the casting is spoiled because of the incorrect position of the core.

(4) In tempering steel or iron, apply too much heat, so that the resulting bars and ingots are of poor quality.

(B) Other Metals

No suggestions available.

PRODUCTION: MINING AND MINERAL EXTRACTION

(A) Coal

(1) A slight blow against your Davy oil lamp will extinguish it, and to light it again you will have to find a place where there is no fire damp. Take a long time looking for the place.

(2) Blacksmiths who make pneumatic picks should not harden them properly, so that they will quickly grow dull.

(3) You can easily put your pneumatic pick out of order. Pour a small amount of water through the oil lever and your pick will stop working. Coal dust and improper lubrication will also put it out of order.

(4) Weaken the chain that pulls the bucket conveyers carrying coal. A deep dent in the chain made with blows of a pick or shovel will cause it to part under normal strain. Once a chain breaks, normally or otherwise take your time about reporting the damage; be slow about taking the chain up for repairs and bringing it back down after repairs.

(5) Derail mine cars by putting obstructions on the rails and in switch points. If possible, pick a gallery where coal cars have to pass each other, so that traffic will be snarled up.

(6) Send up quantities of rock and other useless material with the coal.

PRODUCTION: AGRICULTURE

(A) Machinery

(1) See Industrial Production: Manufacturing, parts (C), (D), and (E), starting on page 21.

(B) Crops and Livestock

Crops and livestock probably will be destroyed only in areas where there are large food surpluses or where the enemy (regime) is known to be requisitioning food.

(1) Feed crops to livestock. Let crops harvest too early or too late. Spoil stores of grain, fruit, and vegetables by soaking them in water so that they will rot. Spoil fruit and vegetables by leaving them in the sun.

TRANSPORTATION: RAILWAYS

(A) Passengers

(1) Make train travel as inconvenient as possible for enemy personnel. Make mistakes in issuing train tickets, leaving portions of the journey uncovered by the ticket book; issue two tickets for the same seat in the train, so that an interesting argument will result; near train time, instead of issuing printed tickets write them out slowly by hand, prolonging the process until the train is nearly ready to leave or has left the station. On station bulletin boards announcing train arrivals and departures, see that false and misleading information is given about trains bound for enemy destinations.

(2) In trains bound for enemy destinations, attendants should make life as uncomfortable as possible for passengers. See that the food is especially bad, take up tickets after midnight, call all station stops very loudly during the night, handle baggage as noisily as possible during the night, and so on.

(3) See that the luggage of enemy personnel is mislaid or unloaded at the wrong stations. Switch address labels on enemy baggage.

(4) Engineers should see that trains run slow or make unscheduled stops for plausible reasons.

(B) Switches, Signals, and Routing

(1) Exchange wires in switchboards containing signals and switches, so that they connect to the wrong terminals.

(2) Loosen push rods so that signal arms do not work; break signal lights; exchange the colored lenses on red and green lights.

(3) Spread and spike switch points in the track so that they will not move, or place rocks or close-packed dirt between the switch points.

(4) Sprinkle rock salt or ordinary salt profusely over the electrical connections of switch points and on the ground nearby. When it rains, the switch will be short-circuited.

(5) See that cars are put on the wrong trains. Remove the labels from cars needing repair and put them on cars in good order. Leave couplings between cars as loose as possible.

(C) Road-beds and Open Track

(1) On a curve, take the bolts out of the tie plates connecting to sections of the outside rail, and scoop away the gravel, cinders, or dirt for a few feet on each side of the connecting joint.

(2) If by disconnecting the tie plate at a joint and loosening sleeper nails on each side of the joint, it becomes possible to move a sections of rail, spread two sections of rail and drive a spike vertically between them.

(D) Oil and Lubrication

(1) See 5 b. **(2)** **(B)**.

(2) Squeeze lubricating pipes with pincers or dent them with hammers, so that the flow of oil is obstructed.

(E) Cooling Systems

(1) See 5 b **(2)** **(C)**.

(F) Gasoline and Oil Fuel

(1) See 5 b **(2)** **(D)**.

(G) Electric Motors

(1) See 5 b **(2)** **(E)** and **(F)**.

(H) Boilers

(1) See 5 b **(2)** **(H)**.

(2) After inspection, put heavy oil or tar in the engines' boilers, or put half a kilogram of soft soap into the water in the tender.

(I) Brakes and Miscellaneous

(1) Engines should run at high speeds and use brakes excessively at curves and on downhill grades.

(2) Punch holes in airbrake valves or water supply pipes.

(3) In the last car of a passenger train or or a front car of a freight, remove the wadding from a journal box and replace it with oily rags.

TRANSPORTATION: AUTOMOTIVE

(A) Roads.

Damage to roads [**(3)** below] is slow, and therefore impractical as a D-day or near D-day activity.

(1) Change signposts at intersections and forks; the enemy will go the wrong way and it may be miles before he discovers his mistakes. In areas where traffic is composed primarily of enemy autos, trucks, and motor convoys of various kinds, remove danger signals from curves and intersections.

(2) When the enemy asks for directions, give him wrong information. Especially when enemy convoys are in the neighborhood, truck drivers can spread rumors and give false information about bridges being out, ferries closed, and detours lying ahead.

(3) If you can start damage to a heavily traveled road, passing traffic and the elements will do the rest. Construction gangs can see that too much sand or water is put in concrete or that the road foundation has soft spots. Anyone can scoop ruts in asphalt and macadam roads, which turn soft in hot weather; passing trucks will accentuate the ruts to a point where substantial repair will bc needed. Dirt roads also can be scooped out. If you are a road laborer, it will be only a few minutcs, work to divert a small stream from a sluice so that it runs over and eats away the road.

(4) Distribute broken glass, nails, and sharp rocks on roads to puncture tires.

(B) Passengers

(1) Bus drivers can go past the stop where the enemy wants to get off. Taxi drivers can waste the enemy's time and make extra money by driving the longest possible route to his destination.

(C) Oil and Lubrication

(1) See 5 b. (2) (B).

(2) Disconnect the oil pump; this will burn out the main bearings in less than 50 miles of normal driving.

(D) Radiator

(1) See 5 b. (2) (C).

(E) Fuel

(1) See 5 b. (2) (D).

(F) Battery and Ignition

(1) Jam bits of wood into the ignition lock; loosen or exchange connections behind the switchboard; put dirt in spark plugs; damage distributor points.

(2) Turn on the lights in parked cars so that the battery will run down.

(3) Mechanics can ruin batteries in a number of undetectable ways: Take the valve cap off a cell, and drive a screw driver slantwise into the exposed water vent, shattering the plates of the cell; no damage will show when you put the cap back on. Iron or copper filings put into the cells (i.e., dropped into the acid) will greatly shorten its life. Copper coins or a few pieces of iron will accomplish the same and more slowly. One hundred to 150 cubic centimeters of vinegar in each cell greatly reduces the life of the battery, but the odor of the vinegar may reveal what has happened.

(G) Gears

(1) Remove the lubricant from or put too light a lubricant in the transmission and other gears.

(2) In trucks, tractors, and other machines with heavy gears, fix the gear case insecurely, putting bolts in only half the bolt holes. The gears will be badly jolted in use and will soon need repairs.

(H) Tires

(1) Slash or puncture tires of unguarded vehicles. Put a nail inside a match box or other small box, and set it vertically in front of the back tire of a stationary car; when the car starts off, the nail will go neatly through the tire.

(2) It is easy to damage a tire in a tire repair shop: In fixing flats, spill glass, benzine, caustic soda, or other material inside the casing, which will puncture or corrode the tube. If you put a gummy substance inside the tube, the next flat will stick the tube to the casing and make it unusable. Or, when you fix a flat tire, you can simply leave between the tube and the casing the object which caused the flat in the first place.

(3) In assembling a tire after repair, pump the tube up as fast as you can. Instead of filling out smoothly, it may crease, in which case it will wear out quickly. Or, as you put a tire together, see if you can pinch the tube between the rim of the tire and the rim of the wheel, so that a blow-out will result.

(4) In putting air into tires, see that they are kept below normal pressure, so that more than an ordinary amount of wear will result. In filling tires on double wheels, inflate the inner tire to a much higher pressure than the outer one; both will wear out more quickly this way. Badly aligned wheels also wear tires out quickly; you can leave wheels out of alignment when they come in for adjustment, or you can spring them out of true with a strong kick, or by driving the car slowly and diagonally into a curb.

(5) If you have access to stocks of tires, you can rot them by spilling oil, gasoline, caustic acid, or benzine on them. Synthetic rubber, however, is less susceptible to these chemicals.

TRANSPORTATION: WATER

(A) Navigation

(1) Barge and river boat personnel should spread false rumors about the navigability and conditions of the waterways they travel. Tell other barge and boat captains to follow channels that will take extra time, or cause them to make canal detours.

(2) Barge and river boat captains should navigate with exceeding caution near locks and bridges, to waste their time and to waste the time of other craft which may have to wait on them. If you don't pump the bilges of ships and barges often enough, they will be slower and harder to navigate. Barges "accidentally" run aground are an efficient time waster too.

(3) Attendants on swing, draw, or bascule bridges can delay traffic over the bridge or in the waterway underneath by being slow. Boat captains can leave unattended draw bridges open in order to hold up road traffic.

(4) Add or subtract compensating magnets to the compass on cargo ships. Demagnetize the compass or maladjust it by concealing a large bar of steel or iron near to it.

(B) Cargo

(1) While loading or unloading, handle cargo carelessly in order to cause damage. Arrange the cargo so that the weakest and lightest crates and boxes will be at the bottom of the hold, while the heaviest ones are on top of them. Put hatch covers and tarpaulins on sloppily, so that rain and deck wash will injure the cargo. Tie float valves open so that storage tanks will overflow on perishable goods.

COMMUNICATIONS

(A) Telephone

(1) At office, hotel, and exchange switchboards, delay putting enemy calls through, give them wrong numbers, cut them off "accidentally," or forget to disconnect them so that the line cannot be used again.

(2) Hamper official and especially military business by making at least one telephone call a day to an enemy headquarters; when you get them, tell them you have the wrong number. Call military or police offices and make anonymous false reports of fires, air raids, or bombs.

(3) In offices and buildings used by the enemy, unscrew the earphone of telephone receivers and remove the diaphragm. Electricians and telephone repair men can make poor connections and damage insulation so that cross talk and other kinds of electrical interference will make conversations hard or impossible to understand.

(4) Put the batteries under automatic switchboards out of commission by dropping nails, metal filings, or coins into the cells. If you can treat half the batteries in this way, the switchboard will stop working. A whole telephone system can be disrupted if you can put 10 percent of the cells in half the batteries of the central battery room out of order.

(B) Telegraph

(1) Delay the transmission and delivery of telegrams to enemy destinations.

(2) Garble telegrams to enemy destinations so that another telegram will have to be sent or a long distance call will have to be made. Sometimes it will be possible to do this by changing a single letter in a word—for example, changing "minimum" to "miximum," so that the person receiving the telegram will not know whether "minimum" or "maximum" is meant.

(C) Transportation Lines

(1) Cut telephone and telegraph transmission lines. Damage insulation on power lines to cause interference.

(D) Mail

(1) Post office employees can see to it that enemy mail is always delayed by one day or more, that it is put in wrong sacks, and so on.

(E) Motion Pictures

(1) Projector operators can ruin newsreels and other enemy propaganda films by bad focusing, speeding up or slowing down the film, and by causing frequent breakage in the film.

(2) Audiences can ruin enemy propaganda films by applauding to drown the words of the speaker, by coughing loudly, and by talking.

(3) Anyone can break up a showing of an enemy propaganda film by putting two or three dozen large moths in a paper bag. Take the bag to the movies with you, put it on the floor in an empty section of the theater as you go in, and leave it open. The moths will fly out and climb into the projector beam, so that the film will be obscured by fluttering shadows.

(F) Radio

(1) Station engineers will find it quite easy to overmodulate transmissions of talks by persons giving enemy propaganda or instructions, so that they will sound as if they were talking through a heavy cotton blanket with a mouth full of marbles.

(2) In your own apartment building, you can interfere with radio reception at times when the enemy wants everybody to listen. Take an electric light plug off the end of an electric light cord; take some wire out of the cord and tie it across two terminals of a two-prong plug or three terminals of a four-prong plug. Then take it around and put it into as many wall and floor outlets as you can find. Each time you insert the plug into a new circuit, you will blow out a fuse and silence all radios running on power from that circuit until a new fuse is put in.

(3) Damaging insulation on any electrical equipment tends to create radio interference in the immediate neighborhood, particularly on large generators, neon signs, fluorescent lighting, X-ray machines, and power lines. If workmen can damage insulation on a high tension line near an enemy airfield, they will make ground-to-plane radio communications difficult and perhaps impossible during long periods of the day.

ELECTRIC POWER

(A) Turbines, Electric Motors, and Transformers

(1) See Industrial Production: Manufacturing (page 21).

(2) See **(E)**, **(F)**,and **(G)**.

(B) Transmission Lines

(1) Linesmen can loosen and dirty insulators to cause power leakage. It will be quite easy, too, for them to tie a piece of very heavy string several times back and forth between two parallel transmission lines, winding it several turns around the wire each time. Beforehand, the string should be heavily saturated with salt and then dried. When it rains, the string becomes a conductor, and a shortcircuit will result.

GENERAL INTERFERENCE WITH ORGANIZATIONS AND PRODUCTION

(A) Organizations and Conferences

(1) Insist on doing everything through "channels." Never permit short-cuts to be taken in order to expedite decisions.

(2) Make "speeches." Talk as frequently as possible and at great length. Illustrate your "points" by long anecdotes and accounts of personal experiences. Never hesitate to make a few appropriate "patriotic" comments.

(3) When possible, refer all matters to committees, for "further study and consideration." Attempt to make the committees as large as possible—never less than five.

(4) Bring up irrelevant issues as frequently as possible.

(5) Haggle over precise wordings of communications, minutes, and resolutions.

(6) Refer back to matters decided upon at the last meeting and attempt to re-open the question of the advisability of that decision.

(7) Advocate "caution." Be "reasonable" and urge your fellow conferees to be "reasonable" and avoid haste, which might result in embarrassments or difficulties later on.

(8) Be worried about the propriety of any decision—raise the question of whether such action as is contemplated lies within the jurisdiction of the group or whether it might conflict with the policy of some higher echelon.

(B) Managers and Supervisors

(1) Demand written orders.

(2) "Misunderstand" orders. Ask endless questions or engage in long correspondence about such orders. Quibble over them when you can.

(3) Do everything possible to delay the delivery of orders. Even though parts of an order may be ready beforehand, don't deliver it until it is completely ready.

(4) Don't order new working materials until your current stocks have been virtually exhausted, so that the slightest delay in filling your order will mean a shutdown.

(5) Order high-quality materials which are hard to get. If you don't get them, argue about it. Warn that inferior materials will mean inferior work.

(6) In making work assignments, always sign out the unimportant jobs first. See that the important jobs are assigned to inefficient workers of poor machines.

(7) Insist on perfect work in relatively unimportant products; send back for refinishing those which have the least flaw. Approve other defective parts whose flaws are not visible to the naked eye.

(8) Make mistakes in routing so that parts and materials will be sent to the wrong place in the plant.

(9) When training new workers, give incomplete or misleading instructions.

(10) To lower morale and with it, production, be pleasant to inefficient workers; give them undeserved promotions. Discriminate against efficient workers; complain unjustly about their work.

(11) Hold conferences when there is more critical work to be done.

(12) Multiply paper work in plausible ways. Start duplicate files.

(13) Multiply the procedures and clearances involved in issuing instructions, paychecks, and so on. See that three people have to approve everything where one would do.

(14) Apply all regulations to the last letter.

(C) Office Workers

(1) Make mistakes in quantities of material when you are copying orders. Confuse similar names. Use wrong addresses.

(2) Prolong correspondence with government bureaus.

(3) Misfile essential documents.

(4) In making carbon copies, make one too few, so that an extra copying job will have to be done.

(5) Tell important callers the boss is busy or talking on another telephone.

(6) Hold up mail until the next collection.

(7) Spread disturbing rumors that sound like inside dope.

(D) Employees

(1) *Work slowly*. Think out ways to increase the number of movements necessary on your job: use a light hammer instead of a heavy one, try to make a small wrench do when a big one is necessary, use little force where considerable force is needed, and so on.

(2) Contrive as many interruptions to your work as you can: when changing the material on which you are working, as you would on a lathe or punch, take needless time to do it. If you are cutting, shaping, or doing other measured work, measure dimensions twice as often as you need to. When you go to the lavatory, spend a longer time there than is necessary. Forget tools so that you will have to go back after them.

(3) Even if you understand the language, pretend not to understand instructions in a foreign tongue.

(4) Pretend that instructions are hard to understand, and ask to have them repeated more than once. Or pretend that you are particularly anxious to do your work, and pester the foreman with unnecessary questions.

(5) Do your work poorly and blame it on bad tools, machinery, or equipment. Complain that these things are preventing you from doing your job right.

(6) Never pass on your skill and experience to a new or less skillful worker.

(7) Snarl up administration in every possible way. Fill out forms illegibly so that they will have to be done over; make mistakes or omit requested information in forms.

(8) If possible, join or help organize a group for presenting employee problems to the management. See that the procedures adopted are as inconvenient as possible for the management, involving the presence of a large number of employees at each presentation, entailing more than one meeting for each grievance, bringing up problems which are largely imaginary, and so on.

(9) Misroute materials.

(10) Mix good parts with unusable scrap and rejected parts.

GENERAL DEVICES FOR LOWERING MORALE AND CREATING CONFUSION

(A) Give lengthy and incomprehensible explanations when questioned.

(B) Report imaginary spies or danger to the Gestapo or police.

(C) Act stupid.

(D) Be as irritable and quarrelsome as possible without getting yourself into trouble.

(E) Misunderstand all sorts of regulations concerning such matters as rationing, transportation, and traffic regulations.

(F) Complain against ersatz materials.

(G) In public, treat Axis nationals or quislings coldly.

(H) Stop all conversation when Axis nationals or quislings enter a cafe.

(I) Cry and sob hysterically at every occasion, especially when confronted by government clerks.

(J) Boycott all movies, entertainments, concerts, and newspapers which are in any way connected with the quisling authorities.

(K) Do not cooperate in salvage schemes.